45 ADDING/SUBTRACTING REAL NUMBERS

ALGEBRA

PROBLEMS

$$5 - [(2 - 3) - 4]$$

WORKBOOK WITH ANSWERS

$$21 - [-(3 - 4) - 2] - 5$$

Najwa Hirn

45 ADDING/SUBTRACTING REAL NUMBERS

ALGEBRA

ADDING/SUBTRACTING REAL NUMBERS

PRACTICE PROBLEMS

WITH ANSWER KEY AND

STEP BY STEP SOLUTIONS

Copyright © 2019 N. Hirn

HOW TO USE THIS WORKBOOK

The study of Mathematics requires understanding of the concepts taught as well as practicing what is learned. It is advisable to practice mathematical problems using pencil and paper to allow the student to follow their train of thoughts as they write each step for the solutions.

This workbook provides 45 problems that require the knowledge and use of the "ADDITION AND SUBTRACTION OF REAL NUMBERS" rules. These rules are summarized in the next section as a refresher for students.

When working out each problem, it is important to learn, understand and apply the rules of "ADDITION AND SUBTRACTION OF REAL NUMBERS" in order to evaluate the mathematical expressions correctly.

The workbook is divided into four sections:

1.   The first section includes a problem per page with ample lined space for the student to solve the problem. Using a pencil will allow the student to erase and redo steps as they become necessary.

2.   The second section is the answer key.

3.   The third section includes detailed solutions for each problem explaining the steps involved in reaching the answer.

Hints are provided when a problem requires the use of concepts that are not part of this workbook. Students should be familiar with some of these concepts.

4.   The final section includes blank worksheets provided for the students to practice additional problems from the assigned text book.

In order to succeed the student should master the material in this section and practice solving each problem.

Have fun as you practice and learn how to **ADD AND SUBTRACT REAL NUMBERS.**

RULES FOR ADDING AND SUBTRACTING REAL NUMBERS

When working on a mathematical expression it is important to understand how to evaluate each expression in order to obtain the correct answer.

In order to learn about "Adding and Subtracting Real numbers", it is important to first understand the following terms:

1. Number Line
2. Real Numbers
3. Absolute value
4. Opposite numbers
5. Terms for Addition and Subtraction

This knowledge can then be used when learning to evaluate the rules associated with "Adding and Subtracting Real numbers" and how to use them to solve a problem.

Explanation

Number Line:

A "Number Line" is a straight line that is drawn and labeled with tick marks. A point **"0"** is labeled somewhere in the middle of the line. The exact location of **"0"** is not important as long as it appears somewhere along the middle of the line.

This point **"0"** is called the **Origin** and it separates the line to a positive side and a negative side.

Tick marks are drawn in both direction of **"0"** at equally spaced distances.

The positive side of the number line is to the right of **"0"**.

The numbers to the right of **"0"** are positive. They increase in value from left to right. Therefore, moving to the right will mean moving in the positive direction.

The numbers to the left of **"0"** are negative. They decrease in value from right to left. Therefore, moving to the left will mean moving in the negative direction.

The following graph illustrates the explanation above:

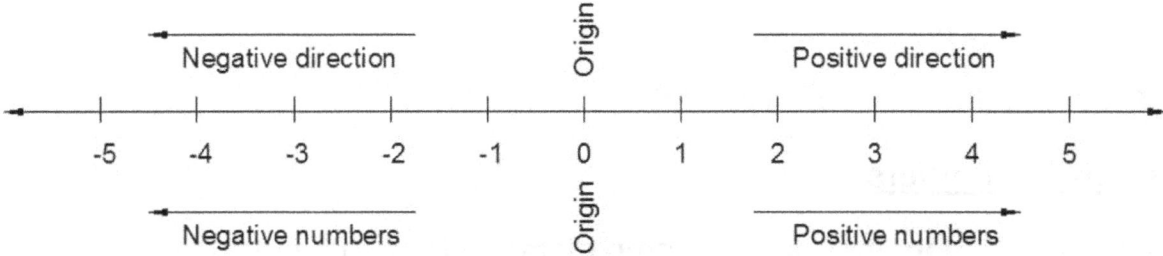

Hint:

When comparing two numbers on the number line, the number on the left is always smaller than the number on the right.

Real Numbers:

Numbers that can be represented with a point on the number line are called "Real Numbers". They include whole numbers, fractions, decimals and others.

Absolute Value:

When a number is placed on the "Number Line" it will have a direction from **"0"** and a distance from **"0"**.

The "Absolute Value" of a real number is its distance from **"0"** and is written enclosed within two short vertical lines as follows: | |

Hint:

*An "Absolute Value" of a number is never negative. It is the distance of that number from **"0"** without regards to what direction it is from **"0"**.*

Example:

The "Absolute Value" of **5** and **−5** both equal **5** and are written as follows:

$$|5| = 5$$

$$|-5| = 5$$

It is just the distance "**5**" from **"0"**.

Opposite Numbers:

They are numbers that are the same distance from **"0"** but are in the opposite direction.

Example:

The opposite number of **5** is **−5** .

The opposite number of **−5** is **5** .

The above definition of opposite numbers applies to all the numbers on the number line.

Terms for Addition and Subtraction

Another term for Addition is "Sum of".

Another term for Subtraction is "Difference of".

Adding and Subtracting Real Numbers:

This workbook will assume that the student is familiar with how to use the Real number line to solve the addition and subtraction of real numbers.

This process, which includes plotting a number on the number line then checking its direction then counting units to arrive at the next number, can become very tedious. It is not functional when trying to solve lengthy mathematical problems and therefore will not be addressed in full within working.

The rules discussed below are derived from the knowledge of what a Real number line represents; the "Absolute value" and "Opposite" of numbers; where to plot numbers on the line and counting their distance from **"0"**.

Rules for Adding and Subtracting Real Numbers:

1.   Adding two numbers with the same sign:

Add their absolute value and use the common sign.

If both numbers are positive the answer will be positive.

If both numbers are negative the answer will be negative.

Example (1):

Adding two positive numbers:

$$2 + 3 = 5$$

The numbers are added, the result is positive.

Example (2):

Adding two negative numbers:

$$-2 - 3 = -5$$

The numbers are added, the result is negative.

Hint:

*When there is no sign in front of a number, that number is always positive. Therefore, the number **2** in the above expression is positive.*

2.   Adding two numbers with different signs:

Subtract the smaller absolute value from the larger.

The answer will have the sign of the number that has the larger absolute value.

Example (1):

$$2 - 3$$

In the above example, the larger number is 3. It has a negative sign in front of it. The absolute value of -3 is 3.

The number 2 is the smaller number. It is positive. Therefore its absolute value is 2.

The solution is reached by subtracting the smaller absolute value from the larger absolute value and using the sign of the larger number in the answer as seen below.

$$2 - 3 = -1$$

Example (2):

$$-2 + 3$$

In the above example, the larger number is 3. It has a positive sign in front of it. The absolute value of 3 is 3.

The number 2 is the smaller number. It is negative. Therefore it the absolute value of -2 is 2.

The solution is reached by subtracting the smaller absolute value from the larger absolute value and using the sign of the larger number in the answer as seen below.

$$-2 + 3 = 1$$

3. Subtracting real numbers.

Subtraction is defined in terms of addition. Therefore, in order to subtract one real number from another, add its' opposite.

The addition rules explained above can then be followed as regards to the signs used for the answers.

Example (1):

Subtract:

$$7 - 2$$

Hint:

Subtracting the 2 is the same as adding −2

The problem can be rewritten as follows:

$$7 + (-2)$$

The addition rule no. (2) explained above is used to solve the problem as follows:

$$7 + (-2) = 5$$

Example (2):

Subtract:

$$-7 - 2$$

Hint:

Subtracting the 2 is the same as adding −2

The problem can be rewritten as follows:

$$-7 + (-2)$$

The addition rule no. (1) explained above is used to solve the problem as follows:

$$-7 + (-2) = -9$$

45 ADDITION/SUBTRACTION PROBLEMS

PROBLEM # 1

SOLVE: $6 + (-3)$

YOUR WORK:

_____

_____

_____

_____

_____

_____

_____

_____

_____

_____

_____

_____

_____

_____

_____

PROBLEM # 2

SOLVE: $13 + (-20)$

YOUR WORK:

_____

_____

_____

_____

_____

_____

_____

_____

_____

_____

_____

_____

_____

_____

_____

_____

PROBLEM # 3

SOLVE: $-7 - 3$

YOUR WORK:

_____

_____

_____

_____

_____

_____

_____

_____

_____

_____

_____

_____

_____

_____

_____

_____

PROBLEM # 4

SOLVE: $-7 - (-3)$

YOUR WORK:

_____

_____

_____

_____

_____

_____

_____

_____

_____

_____

_____

_____

_____

_____

_____

_____

_____

PROBLEM # 5

SOLVE: $-6 + (-6)$

YOUR WORK:

_____

_____

_____

_____

_____

_____

_____

_____

_____

_____

_____

_____

_____

_____

_____

_____

PROBLEM # 6

SOLVE: $-10 + (-15)$

YOUR WORK:

PROBLEM # 7

SOLVE: $-6 + 2$

YOUR WORK:

_____

_____

_____

_____

_____

_____

_____

_____

_____

_____

_____

_____

_____

_____

_____

_____

PROBLEM # 8

SOLVE: $5 + [6 + (-2)] + (-3)$

YOUR WORK:

_____

_____

_____

_____

_____

_____

_____

_____

_____

_____

_____

_____

_____

_____

_____

_____

_____

PROBLEM # 9

SOLVE: $5 - 8$

YOUR WORK:

_____

_____

_____

_____

_____

_____

_____

_____

_____

_____

_____

_____

_____

_____

_____

_____

PROBLEM # 10

SOLVE: $-5 - (2 - 6) - 3$

YOUR WORK:

PROBLEM # 11

SOLVE: $-(2 - 5) - (7 - 3)$

YOUR WORK:

PROBLEM # 12

SOLVE: $-3 + (-2) + [5 + (-4)]$

YOUR WORK:

_____

_____

_____

_____

_____

_____

_____

_____

_____

_____

_____

_____

_____

_____

_____

_____

_____

PROBLEM # 13

SOLVE: $-6 - 9$

YOUR WORK:

_____

_____

_____

_____

_____

_____

_____

_____

_____

_____

_____

_____

_____

_____

_____

PROBLEM # 14

SOLVE: $5 + (-6) + (-7)$

YOUR WORK:

PROBLEM # 15

SOLVE: $(-2 - 5) + (-4)$

YOUR WORK:

_____

_____

_____

_____

_____

_____

_____

_____

_____

_____

_____

_____

_____

_____

PROBLEM # 16

SOLVE: 5 − 5

YOUR WORK:

_____

_____

_____

_____

_____

_____

_____

_____

_____

_____

_____

_____

_____

_____

_____

_____

_____

PROBLEM # 17

SOLVE: $-9 - (-3)$

YOUR WORK:

PROBLEM # 18

SOLVE: *Subtract* 5 *from* − 3

YOUR WORK:

_____

_____

_____

_____

_____

_____

_____

_____

_____

_____

_____

_____

_____

_____

_____

_____

_____

_____

PROBLEM # 19

SOLVE: $[-6 + (-4)] + [7 + (-5)] + (-9)$

YOUR WORK:

PROBLEM # 20

SOLVE: *Find difference of − 4 and 8*

YOUR WORK:

_____

_____

_____

_____

_____

_____

_____

_____

_____

_____

_____

_____

_____

_____

_____

_____

_____

_____

PROBLEM # 21

SOLVE: *Add − 7 to the difference of 2 and 9*

YOUR WORK:

_____

_____

_____

_____

_____

_____

_____

_____

_____

_____

_____

_____

_____

_____

_____

_____

SOLVE: $(-6 + 9) + (-5) + (-4 + 3) + 7$

YOUR WORK:

_____

_____

_____

_____

_____

_____

_____

_____

_____

_____

_____

_____

_____

_____

_____

_____

_____

_____

_____

PROBLEM # 23

SOLVE: $20 + (-6) + [3 + (-9)]$

YOUR WORK:

PROBLEM # 24

SOLVE: *Subtract 4 from − 7*

YOUR WORK:

_____

_____

_____

_____

_____

_____

_____

_____

_____

_____

_____

_____

_____

_____

_____

_____

_____

PROBLEM # 25

SOLVE: $8 - 13$

YOUR WORK:

_____

_____

_____

_____

_____

_____

_____

_____

_____

_____

_____

_____

_____

_____

_____

_____

_____

PROBLEM # 26

SOLVE: $-8 - 13$

YOUR WORK:

_____

_____

_____

_____

_____

_____

_____

_____

_____

_____

_____

_____

_____

_____

_____

_____

_____

PROBLEM # 27

SOLVE: $8 - (-13)$

YOUR WORK:

PROBLEM # 28

SOLVE: $-8 - (-13)$

YOUR WORK:

PROBLEM # 29

SOLVE: *Find the difference of 8 and − 5*

YOUR WORK:

_____

_____

_____

_____

_____

_____

_____

_____

_____

_____

_____

_____

_____

_____

SOLVE: *Subtract 17 from the sum of 4 and* -5

YOUR WORK:

_____

_____

_____

_____

_____

_____

_____

_____

_____

_____

_____

_____

_____

_____

_____

_____

_____

PROBLEM # 31

SOLVE: *subtract 4 from − 3*

YOUR WORK:

_____

_____

_____

_____

_____

_____

_____

_____

_____

_____

_____

_____

_____

_____

PROBLEM # 32

SOLVE: *Subtract 3 from sum of 8 and 1*

YOUR WORK:

_____

_____

_____

_____

_____

_____

_____

_____

_____

_____

_____

_____

_____

_____

_____

_____

_____

PROBLEM # 33

SOLVE: $9 - 2 - 3$

YOUR WORK:

_____

_____

_____

_____

_____

_____

_____

_____

_____

_____

_____

_____

_____

_____

_____

_____

PROBLEM # 34

SOLVE: *Add -4 to the difference of -2 and 5*

YOUR WORK:

_____

_____

_____

_____

_____

_____

_____

_____

_____

_____

_____

_____

_____

_____

_____

_____

_____

_____

PROBLEM # 35

SOLVE: $10 - (-20) - 5$

YOUR WORK:

_____

_____

_____

_____

_____

_____

_____

_____

_____

_____

_____

_____

_____

_____

_____

SOLVE: $-(3 - 10) - (6 - 3)$

YOUR WORK:

PROBLEM # 37

SOLVE: $-6 - 8 - 10$

YOUR WORK:

_____

_____

_____

_____

_____

_____

_____

_____

_____

_____

_____

_____

_____

_____

_____

_____

PROBLEM # 38

SOLVE: $(-9 + 2) + [5 + (-8) + (-4)]$

YOUR WORK:

PROBLEM # 39

SOLVE: $-22 + 4 - 10$

YOUR WORK:

_____

_____

_____

_____

_____

_____

_____

_____

_____

_____

_____

_____

_____

_____

_____

_____

PROBLEM # 40

SOLVE: $5 - [(2 - 3) - 4]$

YOUR WORK:

PROBLEM # 41

SOLVE: $7 - (3 - 9) - 6$

YOUR WORK:

_____

_____

_____

_____

_____

_____

_____

_____

_____

_____

_____

_____

_____

_____

_____

_____

PROBLEM # 42

SOLVE: $[6 + (-2)] + [3 + (-1)]$

YOUR WORK:

PROBLEM # 43

SOLVE: $21 - [-(3 - 4) - 2] - 5$

YOUR WORK:

_____

_____

_____

_____

_____

_____

_____

_____

_____

_____

_____

_____

_____

_____

_____

_____

PROBLEM # 44

SOLVE: $-(5 - 7) - (2 - 8)$

YOUR WORK:

_____

_____

_____

_____

_____

_____

_____

_____

_____

_____

_____

_____

_____

_____

_____

_____

_____

_____

PROBLEM # 45

SOLVE: $16 - [(4 - 5) - 1]$

YOUR WORK:

ANSWER KEY

1. 3

2. -7

3. -10

4. -4

5. -12

6. -25

7. -4

8. 6

9. -3

10. -4

11. -1

12. -4

13. -15

14. -8

15. -11

16. 0

17. -6

18. $-3 - 5 = -8$

19. -17

20. -12

21. -14

22. 4

23. 8

24. $-7 - 4 = -11$

25. -5

26. -21

27. 21

28. 5

29. $8 - (-5) = 13$

30. $[4 + (-5)] - 17 = -18$

31. $-3 - 4 = -7$

32. $(8 + 1) - 3 = 6$

33. 4

34. $(-2 - 5) + (-4) = -11$

35. 25

36. 4

37. -24

38. -14

39. −28

40. 10

41. 7

42. 6

43. 17

44. 8

45. 18

STEP BY STEP SOLUTIONS

PROBLEM # 1

SOLVE: $6 + (-3)$

CORRECT ANSWER: 3

Here is how you solve the problem:

Follow rule (2) that explains how to add two numbers with different signs.

Subtract the smaller absolute value from the larger. Use the sign of the larger number in the answer.

$$6 + (-3) = 3$$

PROBLEM # 2

SOLVE: $13 + (-20)$

CORRECT ANSWER: -7

Here is how you solve the problem:

Follow rule (2) that explains how to add two numbers with different signs.

Subtract the smaller absolute value from the larger. Use the sign of the larger number in the answer.

$$13 + (-20) = -7$$

PROBLEM # 3

SOLVE: $-7 - 3$

CORRECT ANSWER: -10

Here is how you solve the problem:

Follow rule (1) that explains how to add two numbers with the same signs.

Add their absolute value and use the common sign in the answer.

$$-7 - 3 = -10$$

PROBLEM # 4

SOLVE: $-7 - (-3)$

CORRECT ANSWER: -4

Here is how you solve the problem:

Follow rule (3) that explains how to define a subtraction problem in terms of adding the opposite.

Add the opposite.

$$-7 - (-3) =$$

$$-7 + 3$$

Follow rule (2) that explains how to add two numbers with different signs.

Subtract the smaller absolute value from the larger. Use the sign of the larger number in the answer.

$$-7 + 3 = -4$$

PROBLEM # 5

SOLVE: $-6 + (-6)$

CORRECT ANSWER: -12

Here is how you solve the problem:

Follow rule (1) that explains how to add two numbers with the same signs.

Add their absolute value and use the common sign in the answer.

$$-6 + (-6) =$$

$$-6 - 6 = -12$$

PROBLEM # 6

SOLVE: $-10 + (-15)$

CORRECT ANSWER: -25

Here is how you solve the problem:

Follow rule (1) that explains how to add two numbers with the same signs.

Add their absolute value and use the common sign in the answer.

$$-10 + (-15) =$$

$$-10 - 15 = 25$$

PROBLEM # 7

SOLVE: $-6 + 2$

CORRECT ANSWER: -4

Here is how you solve the problem:

Follow rule (2) that explains how to add two numbers with different signs.

Subtract the smaller absolute value from the larger. Use the sign of the larger number in the answer.

$$-6 + 2 = -4$$

SOLVE: $5 + [6 + (-2)] + (-3)$

CORRECT ANSWER: 6

Here is how you solve the problem:

This problem requires following the rules of order of operations **(PEMDAS)** in order to solve.

The rules state that mathematical operations inside parenthesis or brackets are worked first.

It is important to remember that mathematical expressions are worked from inner parenthesis to outer.

Also, mathematical expressions are worked from left to right.

(PEMDAS) rules are explained in the "Order of Operation" workbook which is another part in this series.

Work from left to right and solve the expression inside the bracket first.

Follow rule (2) that explains how to add two numbers with different signs.

Subtract the smaller absolute value from the larger. Use the sign of the larger number in the answer.

$5 + [6 + (-2)] + (-3) =$

$5 + [6 - 2] + (-3) =$

$5 + 4 + (-3)$

Next work left to right and solve the first two expressions.

Follow rule (1) that explains how to add two numbers with the same signs.

Add their absolute value and use the common sign in the answer.

$5 + 4 + (-3) =$

$$9 + (-3)$$

Finally, follow rule (2) that explains how to add two numbers with different signs.

Subtract the smaller absolute value from the larger. Use the sign of the larger number in the answer.

$$9 + (-3) =$$

$$9 - 3 = 6$$

PROBLEM # 9

SOLVE: $5 - 8$

CORRECT ANSWER: -3

Here is how you solve the problem:

Follow rule (3) that explains how to define a subtraction problem in terms of adding the opposite.

Add the opposite.

$$5 + (-8)$$

Follow rule (2) that explains how to add two numbers with different signs.

Subtract the smaller absolute value from the larger. Use the sign of the larger number in the answer.

$$5 + (-8) = -3$$

PROBLEM # 10

SOLVE: $-5 - (2 - 6) - 3$

CORRECT ANSWER: -4

Here is how you solve the problem:

This problem requires following the rules of order of operations **(PEMDAS)** in order to solve.

The rules state that mathematical operations inside parenthesis or brackets are worked first.

It is important to remember that mathematical expressions are worked from inner parenthesis to outer.

Also, mathematical expressions are worked from left to right.

(PEMDAS) rules are explained in the "Order of Operation" workbook which is another part in this series.

Work from left to right and solve the expression inside the parenthesis first.

Follow rule (2) that explains how to add two numbers with different signs.

Subtract the smaller absolute value from the larger. Use the sign of the larger number in the answer.

$-5 - (2 - 6) - 3 =$

$-5 - (-4) - 3$

Work with the first two expressions and open the parenthesis. Follow rule (3) that explains how to define a subtraction problem in terms of adding the opposite.

Add the opposite.

$-5 + 4 - 3$

Next, follow rule (2) that explains how to add two numbers with different signs.

Subtract the smaller absolute value from the larger. Use the sign of the larger number in the answer.

$-1 - 3$

Follow rule (3) that explains how to define a subtraction problem in terms of adding the opposite.

Add the opposite.

$-1 + (-3)$

Finally, follow rule (1) that explains how to add two numbers with the same signs.

Add their absolute value and use the common sign in the answer.

$-1 - 3 = -4$

PROBLEM # 11

SOLVE: $-(2 - 5) - (7 - 3)$

CORRECT ANSWER: −1

Here is how you solve the problem:

This problem requires following the rules of order of operations **(PEMDAS)** in order to solve.

The rules state that mathematical operations inside parenthesis or brackets are worked first.

It is important to remember that mathematical expressions are worked from inner parenthesis to outer.

Also, mathematical expressions are worked from left to right.

(PEMDAS) rules are explained in the "Order of Operation" workbook which is another part in this series.

Rewrite the expressions inside the parenthesis as follows:

$$-(+2-5)-(+7-3)$$

Both expressions inside each parenthesis follow rule (2) that explains how to add two numbers with different signs.

Subtract the smaller absolute value from the larger. Use the sign of the larger number in the answer.

$$-(+2-5)-(+7-3)=$$

$$-(-3)-(+4)$$

Follow rule (3) that explains how to define a subtraction problem in terms of adding the opposite.

Add the opposite.

$$+3-4$$

Follow rule (2) that explains how to add two numbers with different signs.

Subtract the smaller absolute value from the larger. Use the sign of the larger number in the answer.

$$+3-4=-1$$

PROBLEM # 12

SOLVE: $-3+(-2)+[5+(-4)]$

CORRECT ANSWER: -4

Here is how you solve the problem:

This problem requires following the rules of order of operations **(PEMDAS)** in order to solve.

The rules state that mathematical operations inside parenthesis or brackets are worked first.

It is important to remember that mathematical expressions are worked from inner parenthesis to outer.

Also, mathematical expressions are worked from left to right.

(PEMDAS) rules are explained in the "Order of Operation" workbook which is another part in this series.

Work from left to right and solve the expression inside the bracket first.

Follow rule (2) that explains how to add two numbers with different signs.

Subtract the smaller absolute value from the larger. Use the sign of the larger number in the answer.

$$-3 + (-2) + [5 - 4] =$$

$$-3 + (-2) + 1$$

Work from left to right and solve the first two expressions first.

Follow rule (1) that explains how to add two numbers with the same signs.

Add their absolute value and use the common sign in the answer.

$$-3 + (-2) + 1 =$$

$$-5 + 1$$

Finally, follow rule (2) that explains how to add two numbers with different signs.

Subtract the smaller absolute value from the larger. Use the sign of the larger number in the answer.

$$-5 + 1 = -4$$

PROBLEM # 13

SOLVE: $-6 - 9$

CORRECT ANSWER: -15

Here is how you solve the problem:

Follow rule (1) that explains how to add two numbers with the same signs.

Add their absolute value and use the common sign in the answer.

$-6 + (-9) = -15$

PROBLEM # 14

SOLVE: $5 + (-6) + (-7)$

CORRECT ANSWER: -8

Here is how you solve the problem:

Mathematical expressions are worked from left to right.

Start with the first two expressions and follow rule (2) that explains how to add two numbers with different signs.

Subtract the smaller absolute value from the larger. Use the sign of the larger number in the answer.

$5 + (-6) + (-7) =$

$-1 + (-7)$

Follow rule (1) that explains how to add two numbers with the same signs.

Add their absolute value and use the common sign in the answer.

$-1 - 7 = -8$

PROBLEM # 15

SOLVE: $(-2 - 5) + (-4)$

CORRECT ANSWER: -11

Here is how you solve the problem:

This problem requires following the rules of order of operations **(PEMDAS)** in order to solve.

The rules state that mathematical operations inside parenthesis or brackets are worked first.

It is important to remember that mathematical expressions are worked from inner parenthesis to outer.

Also, mathematical expressions are worked from left to right.

(PEMDAS) rules are explained in the "Order of Operation" workbook which is another part in this series.

Work from left to right and solve the expression inside the parenthesis first.

Follows rule (1) that explains how to add two numbers with the same signs.

Add their absolute value and use the common sign in the answer.

$-2 - 5 + (-4) =$

$-7 + (-4)$

Follows rule (1) that explains how to add two numbers with the same signs.

Add their absolute value and use the common sign in the answer.

$-7 - 4 = -11$

PROBLEM # 16

SOLVE: $5 - 5$

CORRECT ANSWER: 0

Here is how you solve the problem:

Follow rule (2) that explains how to add two numbers with different signs.

Subtract the smaller absolute value from the larger. Use the sign of the larger number in the answer.

$5 - 5 =$

$+5 - 5 = 0$

PROBLEM # 17

SOLVE: $-9 - (-3)$

CORRECT ANSWER: −6

Here is how you solve the problem:

Follow rule (3) that explains how to define a subtraction problem in terms of adding the opposite.

Add the opposite.

$-9 - (-3) =$

$-9 + 3$

Follow rule (2) that explains how to add two numbers with different signs.

Subtract the smaller absolute value from the larger. Use the sign of the larger number in the answer.

$-9 + 3 = -6$

PROBLEM # 18

SOLVE: *Subtract* **5** *from* **− 3**

CORRECT ANSWER: -8

Here is how you solve the problem:

Write the problem in a mathematical expression.

$-3 - 5$

Follow rule (1) that explains how to add two numbers with the same signs.

Add their absolute value and use the common sign in the answer.

$-3 - 5 = -8$

PROBLEM # 19

SOLVE: $[-6 + (-4)] + [7 + (-5)] + (-9)$

CORRECT ANSWER: -17

Here is how you solve the problem:

This problem requires following the rules of order of operations **(PEMDAS)** in order to solve.

The rules state that mathematical operations inside parenthesis or brackets are worked first.

It is important to remember that mathematical expressions are worked from inner parenthesis to outer.

Also, mathematical expressions are worked from left to right.

(PEMDAS) rules are explained in the "Order of Operation" workbook which is another part in this series.

Working from left to right solve what is inside the first bracket that contains an addition/subtraction operation.

Follow rule (1) that explains how to add two numbers with the same signs.

Add their absolute value and use the common sign in the answer.

$$-6 - 4 + [7 + (-5)] + (-9) =$$

$$-10 + [7 + (-5)] + (-9)$$

Next work what is inside the second bracket that contains an addition/ subtraction operation.

Follow rule (2) that explains how to add two numbers with different signs.

Subtract the smaller absolute value from the larger. Use the sign of the larger number in the answer.

$$-10 + [7 + (-5)] + (-9) =$$

$$-10 + [7 - 5] + (-9) =$$

$$-10 + 2 + (-9)$$

Next open the last parenthesis as follows:

$$-10 + 2 - 9$$

Work from left to right to solve the first two expressions.

Follow rule (2) that explains how to add two numbers with different signs.

Subtract the smaller absolute value from the larger. Use the sign of the larger number in the answer.

$$-10 + 2 - 9 =$$

$$-8 - 9$$

Finally, follow rule (1) that explains how to add two numbers with the same signs.

Add their absolute value and use the common sign in the answer.

$-8 - 9 = -17$

PROBLEM # 20

SOLVE: *Find the difference of* -4 *and* 8

CORRECT ANSWER: -12

Here is how you solve the problem:

Write the problem in a mathematical expression.

Hint:

The word "Difference" means subtraction.

$-4 - 8$

Follow rule (1) that explains how to add two numbers with the same signs.

Add their absolute value and use the common sign in the answer.

$-4 - 8 = -12$

PROBLEM # 21

SOLVE: *Add* -7 *to the difference of* 2 *and* 9

CORRECT ANSWER: -14

Here is how you solve the problem:

Write the problem in a mathematical expression.

Hint:

The word "Difference" means subtraction.

$(2 - 9) + (-7)$

This problem requires following the rules of order of operations **(PEMDAS)** in order to solve.

The rules state that mathematical operations inside parenthesis or brackets are worked first.

It is important to remember that mathematical expressions are worked from inner parenthesis to outer.

Also, mathematical expressions are worked from left to right.

(PEMDAS) rules are explained in the "Order of Operation" workbook which is another part in this series.

Work from left to right and solve the expression inside the first parenthesis first.

Follow rule (2) that explains how to add two numbers with different signs.

Subtract the smaller absolute value from the larger. Use the sign of the larger number in the answer.

$(+2 - 9) + (-7) =$

$-7 - 7$

Follow rule (1) that explains how to add two numbers with the same signs.

Add their absolute value and use the common sign in the answer.

$-7 - 7 = -14$

PROBLEM # 22

SOLVE: $(-6 + 9) + (-5) + (-4 + 3) + 7$

CORRECT ANSWER: 4

Here is how you solve the problem:

This problem requires following the rules of order of operations **(PEMDAS)** in order to solve.

The rules state that mathematical operations inside parenthesis or brackets are worked first.

It is important to remember that mathematical expressions are worked from inner parenthesis to outer.

Also, mathematical expressions are worked from left to right.

(PEMDAS) rules are explained in the "Order of Operation" workbook which is another part in this series.

Working from left to right solve what is inside the first parenthesis that contains the addition/subtraction operation.

Follow rule (2) that explains how to add two numbers with different signs.

Subtract the smaller absolute value from the larger. Use the sign of the larger number in the answer.

$$(-6 + 9) + (-5) + (-4 + 3) + 7 =$$

$$3 + (-5) + (-4 + 3) + 7$$

Next, work from left to right solve what is inside the second parenthesis that includes the addition/subtraction operation.

Follow rule (2) that explains how to add two numbers with different signs.

Subtract the smaller absolute value from the larger. Use the sign of the larger number in the answer.

$$3 + (-5) + (-4 + 3) + 7 =$$

$$3 + (-5) + (-1) + 7 =$$

$$3 - 5 - 1 + 7$$

Work from left to right to solve the first two expressions.

Follow rule (2) that explains how to add two numbers with different signs.

Subtract the smaller absolute value from the larger. Use the sign of the larger number in the answer.

$+3 - 5 - 1 + 7 =$

$-2 - 1 + 7$

Next, work from left to right again to solve the first two expressions.

Follow rule (1) that explains how to add two numbers with the same signs.

Add their absolute value and use the common sign in the answer.

$-2 - 1 + 7 =$

$-3 + 7 =$

Finally, follow rule (2) that explains how to add two numbers with different signs.

Subtract the smaller absolute value from the larger. Use the sign of the larger number in the answer.

$-3 + 7 = 4$

PROBLEM # 23

SOLVE: $20 + (-6) + [3 + (-9)]$

CORRECT ANSWER: 8

Here is how you solve the problem:

This problem requires following the rules of order of operations **(PEMDAS)** in order to solve.

The rules state that mathematical operations inside parenthesis or brackets are worked first.

It is important to remember that mathematical expressions are worked from inner parenthesis to outer.

Also, mathematical expressions are worked from left to right.

(PEMDAS) rules are explained in the "Order of Operation" workbook which is another part in this series.

Working from left to right solve what is inside the bracket that contains the addition/subtraction operation.

Follow rule (2) that explains how to add two numbers with different signs.

Subtract the smaller absolute value from the larger. Use the sign of the larger number in the answer.

$$20 + (-6) + [3 + (-9)] =$$

$$20 + (-6) + [3 - 9] =$$

$$20 + (-6) + (-6)$$

Work from left to right solve the first two expressions.

Follow rule (2) that explains how to add two numbers with different signs.

Subtract the smaller absolute value from the larger. Use the sign of the larger number in the answer.

$$20 - 6 + (-6) =$$

$$14 + (-6)$$

Next, work from left to right again to solve the expression.

Follow rule (2) that explains how to add two numbers with different signs.

Subtract the smaller absolute value from the larger. Use the sign of the larger number in the answer.

$$+14 - 6 = 8$$

PROBLEM # 24

SOLVE: *Subtract 4 from −7*

CORRECT ANSWER: -11

Here is how you solve the problem:

Write the problem in a mathematical expression.

$$-7 - 4$$

Follow rule (1) that explains how to add two numbers with the same signs.

Add their absolute value and use the common sign in the answer.

$$-7 - 4 = -11$$

PROBLEM # 25

SOLVE: $8 - 13$

CORRECT ANSWER: -5

Here is how you solve the problem:

Follow rule (2) that explains how to add two numbers with different signs.

Subtract the smaller absolute value from the larger. Use the sign of the larger number in the answer.

$$+8 - 13 = -5$$

PROBLEM # 26

SOLVE: $-8 - 13$

CORRECT ANSWER: -21

Here is how you solve the problem:

Follow rule (1) that explains how to add two numbers with the same signs.

Add their absolute value and use the common sign in the answer.

$-8 - 13 = -21$

PROBLEM # 27

SOLVE: $8 - (-13)$

CORRECT ANSWER: 21

Here is how you solve the problem:

Follow rule (3) that explains how to define a subtraction problem in terms of adding the opposite.

Add the opposite.

$8 + 13$

Follow rule (1) that explains how to add two numbers with the same signs.

Add their absolute value and use the common sign in the answer.

$8 + 13 = 21$

PROBLEM # 28

SOLVE: $-8 - (-13)$

CORRECT ANSWER: 5

Here is how you solve the problem:

First follow rule (3) that explains how to define a subtraction problem in terms of adding the opposite as follows:

Add the opposite.

$-8 + 13$

Next, follow rule (2) that explains how to add two numbers with different signs.

Subtract the smaller absolute value from the larger. Use the sign of the larger number in the answer.

$-8 + 13 = 5$

PROBLEM # 29

SOLVE: *Find the difference of 8 and* -5

CORRECT ANSWER: 13

Here is how you solve the problem:

Write the problem in a mathematical expression.

Hint:

The word "Difference" means subtraction.

$8 - (-5)$

Next, Follow rule (3) that explains how to define a subtraction problem in terms of adding the opposite.

Add the opposite.

$$8 + 5$$

Finally, follow rule (1) that explains how to add two numbers with the same signs.

Add their absolute value and use the common sign in the answer.

$$8 + 5 = 13$$

PROBLEM # 30

SOLVE: *Subtract 17 from the sum of 4 and − 5*

CORRECT ANSWER: −18

Here is how you solve the problem:

Write the problem in a mathematical expression.

$$[4 + (-5)] - 17$$

This problem requires following the rules of order of operations **(PEMDAS)** in order to solve.

The rules state that mathematical operations inside parenthesis or brackets are worked first.

It is important to remember that mathematical expressions are worked from inner parenthesis to outer.

Also, mathematical expressions are worked from left to right.

(PEMDAS) rules are explained in the "Order of Operation" workbook which is another part in this series.

Work from left to right and solve what is inside the bracket first.

Follow rule (2) that explains how to add two numbers with different signs.

Subtract the smaller absolute value from the larger. Use the sign of the larger number in the answer.

$$[4 - 5] - 17 =$$

$$[-1] - 17$$

Finally, follow rule (1) that explains how to add two numbers with the same signs.

Add their absolute value and use the common sign in the answer.

$$-1 - 17 = -18$$

PROBLEM # 31

SOLVE: $subtract\ 4\ from - 3$

CORRECT ANSWER: -7

Here is how you solve the problem:

Write the problem in a mathematical expression.

$$-3 - 4$$

Finally, follow rule (1) that explains how to add two numbers with the same signs.

Add their absolute value and use the common sign in the answer.

$$-3 - 4 = -7$$

PROBLEM # 32

SOLVE: *Subtract 3 from the sum of 8 and 1*

CORRECT ANSWER: 6

Here is how you solve the problem:

Write the problem in a mathematical expression.

$(8 + 1) - 3$

This problem requires following the rules of order of operations **(PEMDAS)** in order to solve.

The rules state that mathematical operations inside parenthesis or brackets are worked first.

It is important to remember that mathematical expressions are worked from inner parenthesis to outer.

Also, mathematical expressions are worked from left to right.

(PEMDAS) rules are explained in the "Order of Operation" workbook which is another part in this series.

Work from left to right and solve the expression inside the parenthesis first.

Follow rule (1) that explains how to add two numbers with the same signs.

Add their absolute value and use the common sign in the answer.

$(8 + 1) - 3 =$

$9 - 3$

Finally, follow rule (2) that explains how to add two numbers with different signs.

Subtract the smaller absolute value from the larger. Use the sign of the larger number in the answer.

$9 - 3 = 6$

PROBLEM # 33

SOLVE: $9 - 2 - 3$

CORRECT ANSWER: 4

Here is how you solve the problem:

Working from left to right solve the first two expressions.

Follow rule (2) that explains how to add two numbers with different signs.

Subtract the smaller absolute value from the larger. Use the sign of the larger number in the answer.

$9 - 2 - 3 =$

$7 - 3$

Finally, follow rule (2) that explains how to add two numbers with different signs.

Subtract the smaller absolute value from the larger. Use the sign of the larger number for the answer.

$7 - 3 = 4$

PROBLEM # 34

SOLVE: *Add* $- 4$ *to the difference of* $- 2$

CORRECT ANSWER: -11

Here is how you solve the problem:

Write the problem in a mathematical expression.

The word "Difference" means subtraction.

$$(-2 - 5) + (-4)$$

This problem requires following the rules of order of operations **(PEMDAS)** in order to solve.

The rules state that mathematical operations inside parenthesis or brackets are worked first.

It is important to remember that mathematical expressions are worked from inner parenthesis to outer.

Also, mathematical expressions are worked from left to right.

(PEMDAS) rules are explained in the "Order of Operation" workbook which is another part in this series.

Work from left to right and solve the expression inside the parenthesis first.

Follow rule (1) that explains how to add two numbers with the same signs.

Add their absolute value and use the common sign in the answer.

$$(-2 - 5) + (-4) =$$

$$-7 + (-4)$$

In order to complete the problem, follow rule (1) again that explains how to add two numbers with the same signs.

Add their absolute value and use the common sign in the answer.

$$-7 + (-4) =$$

$$-7 - 4 = -11$$

PROBLEM # 35

SOLVE: $10 - (-20) - 5$

CORRECT ANSWER: 25

Here is how you solve the problem:

Working from left to right solve the first two expressions.

Follow rule (3) that explains how to define a subtraction problem in terms of adding the opposite.

Add the opposite.

$10 - (-20) - 5 =$

$10 + 20 - 5$

Follow rule (1) that explains how to add two numbers with the same signs.

Add their absolute value and use the common sign in the answer.

$10 + 20 - 5 =$

$30 - 5$

Follow rule (3) that explains how to define a subtraction problem in terms of adding the opposite.

$30 - 5 =$

$30 + (-5)$

Follow rule (2) that explains how to add two numbers with different signs.

Subtract the smaller absolute value from the larger. Use the sign of the larger for number for the answer.

$30 + (-5) = 25$

SOLVE: $-(3 - 10) - (6 - 3)$

CORRECT ANSWER: 4

This problem requires following the rules of order of operations **(PEMDAS)** in order to solve.

The rules state that mathematical operations inside parenthesis or brackets are worked first.

It is important to remember that mathematical expressions are worked from inner parenthesis to outer.

Also, mathematical expressions are worked from left to right.

(PEMDAS) rules are explained in the "Order of Operation" workbook which is another part in this series.

Working from left to right solve the expression inside the first parenthesis.

Follow rule (2) that explains how to add two numbers with different signs.

Subtract the smaller absolute value from the larger. Use the sign of the larger number in the answer.

$-(+3 - 10) - (6 - 3) =$

$-(-7) - (6 - 3)$

Next, solve the expression inside the second parenthesis.

Follow rule (2) that explains how to add two numbers with different signs.

Subtract the smaller absolute value from the larger. Use the sign of the larger number in the answer.

$-(-7) - (6 - 3) =$

$-(-7) - (+6 - 3) =$

$-(-7) - (3) =$

Follow rule (3) that explains how to define a subtraction problem in terms of adding the opposite.

Add the opposite.

$-(-7) - (3) =$

$+7 - 3$

Finally, follow rule (2) that explains how to add two numbers with different signs.

Subtract the smaller absolute value from the larger. Use the sign of the larger number in the answer.

$7 - 3 = 4$

PROBLEM # 37

SOLVE: $-6 - 8 - 10$

CORRECT ANSWER: -24

Here is how you solve the problem:

Working from left to right solve the first two expressions.

Follow rule (1) that explains how to add two numbers with the same signs.

Add their absolute value and use the common sign in the answer.

$-6 - 8 - 10 =$

$-14 - 10$

Finally, follow rule (1) that explains how to add two numbers with the same signs.

Add their absolute value and use the common sign in the answer.

$$-14 - 10 = -24$$

PROBLEM # 38

SOLVE: $(-9 + 2) + [5 + (-8) + (-4)]$

CORRECT ANSWER: -14

Here is how you solve the problem:

This problem requires following the rules of order of operations **(PEMDAS)** in order to solve.

The rules state that mathematical operations inside parenthesis or brackets are worked first.

It is important to remember that mathematical expressions are worked from inner parenthesis to outer.

Also, mathematical expressions are worked from left to right.

(PEMDAS) rules are explained in the "Order of Operation" workbook which is another part in this series.

Work from left to right and solve the expression inside the first parenthesis.

Follow rule (2) that explains how to add two numbers with different signs.

Subtract the smaller absolute value from the larger. Use the sign of the larger number in the answer.

$$(-9 + 2) + [5 + (-8) + (-4)] =$$

$$-7 + [5 + (-8) + (-4)]$$

Next solve the first two expressions inside the bracket.

Follow rule (2) that explains how to add two numbers with different signs.

Subtract the smaller absolute value from the larger. Use the sign of the larger number in the answer.

$$-7 + [5 + (-8) + (-4)] =$$

$$-7 + [5 - 8 + (-4)] =$$

$$-7 + [-3 + (-4)] =$$

Next solve the expression inside the bracket.

Follow rule (1) that explains how to add two numbers with the same signs.

Add their absolute value and use the common sign in the answer.

$$-7 + [-3 + (-4)] =$$

$$-7 + [-3 - 4] =$$

$$-7 + [-7]$$

Finally, follow rule (1) that explains how to add two numbers with the same signs.

Add their absolute value and use the common sign in the answer.

$$-7 + [-7] =$$

$$-7 - 7 = -14$$

PROBLEM # 39

SOLVE: $-22 + 4 - 10$

CORRECT ANSWER: -28

Here is how you solve the problem:

Working from left to right solve the first two expressions.

Follow rule (2) that explains how to add two numbers with different signs.

Subtract the smaller absolute value from the larger. Use the sign of the larger number in the answer.

$$-22 + 4 - 10 =$$

$$-18 - 10$$

Finally, follow rule (1) that explains how to add two numbers with the same signs.

Add their absolute value and use the common sign in the answer.

$$-18 - 10 = -28$$

PROBLEM # 40

SOLVE: $5 - [(2 - 3) - 4]$

CORRECT ANSWER: 10

Here is how you solve the problem:

This problem requires following the rules of order of operations **(PEMDAS)** in order to solve.

The rules state that mathematical operations inside parenthesis or brackets are worked first.

It is important to remember that mathematical expressions are worked from inner parenthesis to outer.

Also, mathematical expressions are worked from left to right.

(PEMDAS) rules are explained in the "Order of Operation" workbook which is another part in this series.

Work from left to right and solve what is inside the bracket, starting by solving the parenthesis first as follows:

For the expression inside the inner parenthesis follow rule (2) that explains how to add two numbers with different signs.

Subtract the smaller absolute value from the larger. Use the sign of the larger number in the answer.

$$5 - [(2 - 3) - 4 =$$

$$5 - [-1 - 4]$$

For the expression inside the bracket follow rule (1) that explains how to add two numbers with the same signs.

Add their absolute value and use the common sign in the answer.

$$5 - [-1 - 4] =$$

$$5 - [-5]$$

Follow rule (3) that explains how to define a subtraction problem in terms of adding the opposite.

Add the opposite.

$$5 - [-5] =$$

$$5 + 5$$

Finally, follow rule (1) that explains how to add two numbers with the same signs.

Add their absolute value and use the common sign in the answer.

$$5 + 5 = 10$$

PROBLEM # 41

SOLVE: $7 - (3 - 9) - 6$

CORRECT ANSWER: 7

Here is how you solve the problem:

This problem requires following the rules of order of operations **(PEMDAS)** in order to solve.

The rules state that mathematical operations inside parenthesis or brackets are worked first.

It is important to remember that mathematical expressions are worked from inner parenthesis to outer.

Also, mathematical expressions are worked from left to right.

(PEMDAS) rules are explained in the "Order of Operation" workbook which is another part in this series.

Work from left to right and solve what is inside the parenthesis .

For the expression inside the inner parenthesis follow rule (2) that explains how to add two numbers with different signs.

Subtract the smaller absolute value from the larger. Use the sign of the larger number in the answer.

$7 - (3 - 9) - 6 =$

$7 - (-6) - 6$

Work from left to right and solve the first two expressions.

In order to solve the first two expressions follow rule (3) that explains how to define a subtraction problem in terms of adding the opposite.

Add the opposite.

$7 - (-6) - 6 =$

$7 + 6 - 6$

The first two expressions now follow rule (1) that explains how to add two numbers with the same signs.

Add their absolute value and use the common sign in the answer.

$7 + 6 - 6 =$

$13 - 6$

Finally, follow rule (2) that explains how to add two numbers with different signs.

Subtract the smaller absolute value from the larger. Use the sign of the larger number in the answer.

$13 - 6 = 7$

PROBLEM # 42

SOLVE: $[6 + (-2)] + [3 + (-1)]$

CORRECT ANSWER: 6

Here is how you solve the problem:

This problem requires following the rules of order of operations **(PEMDAS)** in order to solve.

The rules state that mathematical operations inside parenthesis or brackets are worked first.

It is important to remember that mathematical expressions are worked from inner parenthesis to outer.

Also, mathematical expressions are worked from left to right.

(PEMDAS) rules are explained in the "Order of Operation" workbook which is another part in this series.

Work from left to right and solve what is inside each bracket.

Notice that the expressions inside both brackets follow rule (2) that explains how to add two numbers with different signs.

Subtract the smaller absolute value from the larger. Use the sign of the larger number in the answer.

$$[6 + (-2)] + [3 + (-1)] =$$

$$[6 - 2] + [3 - 1] =$$

$$4 + 2$$

Finally, follow rule (1) that explains how to add two numbers with the same signs.

Add their absolute value and use the common sign in the answer.

$$4 + 2 = 6$$

PROBLEM # 43

SOLVE: $21 - [-(3 - 4) - 2] - 5$

CORRECT ANSWER: 17

Here is how you solve the problem:

This problem requires following the rules of order of operations **(PEMDAS)** in order to solve.

The rules state that mathematical operations inside parenthesis or brackets are worked first.

It is important to remember that mathematical expressions are worked from inner parenthesis to outer.

Also, mathematical expressions are worked from left to right.

(PEMDAS) rules are explained in the "Order of Operation" workbook which is another part in this series.

Work from left to right and solve what is inside the inner parenthesis first.

Notice that the expression inside the inner parenthesis follows rule (2) that explains how to add two numbers with different signs.

Subtract the smaller absolute value from the larger. Use the sign of the larger number in the answer.

$$21 - [-(3 - 4) - 2] - 5 =$$

$$21 - [-(-1) - 2] - 5$$

Next work inside the bracket.

Follow rule (3) that explains how to define a subtraction problem in terms of adding the opposite.

Add the opposite.

$$21 - [+1 - 2] - 5$$

Continue working inside the bracket.

Follows rule (2) that explains how to add two numbers with different signs.

Subtract the smaller absolute value from the larger. Use the sign of the larger number in the answer.

$$21 - [+1 - 2] - 5 =$$

$$21 - [-1] - 5$$

The first two expressions now follow rule (3) that explains how to define a subtraction problem in terms of adding the opposite.

Add the opposite.

$$21 + 1 - 5$$

Work from left to right and solve the first two expressions.

Follow rule (1) that explains how to add two numbers with the same signs.

Add their absolute value and use the common sign in the answer.

$$21 + 1 - 5 =$$

$$22 - 5$$

Finally, follows rule (2) that explains how to add two numbers with different signs.

Subtract the smaller absolute value from the larger. Use the sign of the larger number in the answer.

$$22 - 5 = 17$$

PROBLEM # 44

SOLVE: $-(5 - 7) - (2 - 8)$

CORRECT ANSWER: 8

Here is how you solve the problem:

This problem requires following the rules of order of operations **(PEMDAS)** in order to solve.

The rules state that mathematical operations inside parenthesis or brackets are worked first.

It is important to remember that mathematical expressions are worked from inner parenthesis to outer.

Also, mathematical expressions are worked from left to right.

(PEMDAS) rules are explained in the "Order of Operation" workbook which is another part in this series.

Work from left to right and solve what is inside each parenthesis.

For each expression inside each parenthesis, follows rule (2) that explains how to add two numbers with different signs.

Subtract the smaller absolute value from the larger. Use the sign of the larger number in the answer.

$$-(-2) - (-6)$$

Next, follow rule (3) that explains how to define a subtraction problem in terms of adding the opposite.

Add the opposite.

$$+2 + 6 =$$

Follow rule (1) that explains how to add two numbers with the same signs.

Add their absolute value and use the common sign in the answer.

$$+2 + 6 = 8$$

PROBLEM # 45

SOLVE: $16 - [(4 - 5) - 1]$

CORRECT ANSWER: 18

Here is how you solve the problem:

This problem requires following the rules of order of operations **(PEMDAS)** in order to solve.

The rules state that mathematical operations inside parenthesis or brackets are worked first.

It is important to remember that mathematical expressions are worked from inner parenthesis to outer.

Also, mathematical expressions are worked from left to right.

(PEMDAS) rules are explained in the "Order of Operation" workbook which is another part in this series.

Work from left to right and solve the expression inside the inner parenthesis first.

Follows rule (2) that explains how to add two numbers with different signs.

Subtract the smaller absolute value from the larger. Use the sign of the

larger number in the answer.

$$16 - [(4 - 5) - 1] =$$

$$16 - [-1 - 1]$$

Work from left to right and solve the expression inside the bracket.

Follow rule (1) that explains how to add two numbers with the same signs.

Add their absolute value and use the common sign in the answer.

$$16 - [-1 - 1] =$$

$$16 - [-2]$$

Follow rule (3) that explains how to define a subtraction problem in terms of adding the opposite.

Add the opposite.

$$16 - [-2] =$$

$$16 + 2$$

Finally, follow rule (1) that explains how to add two numbers with the same signs.

Add their absolute value and use the common sign in the answer.

$$16 + 2 = 18$$

BLANK WORKSHEETS

ABOUT THE AUTHOR

Najwa Hirn holds a Bachelors of Science degree with honors in Engineering Technology. She has been working with Mathematics for over 25 years both professionally and privately. She taught math for many years.

Najwa is passionate about helping students succeed in Mathematics. She prides herself in being able to simplify math concept for students and teach every them according to their levels. Her step by step approach to solving problems has helped many students understand concepts better. She does not eliminate a step no matter how simple it may be since eliminating steps is what confuses many students.